ABOUT THIS GUIDE

Waterfowl can be identified by using information about size, shape, color pattern, markings, behavior, habitat, range, and calls. All clues are important, but by far the most important things to concentrate on for this group are shape and patterns of white.

This guide features more than 25 species of sea ducks, geese, loons, grebes, and other waterfowl. It includes silhouettes that emphasize distinctive shape and placement of white, and full color images of pairs and seasonal changes. Sections address behavior, habitat, color patterns, and other characteristics important for each species.

ABOUT THE AUTHOR

Kevin J. McGowan, project manager for Distance Learning in Bird Biology at the Cornell Lab, developed the Cornell Lab of Ornithology Waterfowl ID System, "Where's the White?" based on years of experience as an ornithologist and avid birder. He has also studied crows in central New York since 1988, following the life stories of more than 2,000 banded birds. His published work includes scientific papers, *The Second Atlas of Breeding Birds in New York State*, educational websites and online courses, and a series of webinars.

ONLINE EDUCATION FROM THE CORNELL LAB

For more than 45 years the Cornell Lab has pioneered distance learning – everything from a correspondence course in bird biology to real-time webinars on topics of interest to people who care about birds. Learn more about birds by joining our course on Bird Academy, Be a Better Birder: Ducks and Waterfowl Identification. Visit **academy.allaboutbirds.org**.

THE CORNELL LAB OF ORNITHOLOGY is a world leader in the study, appreciation, and conservation of birds. Our hallmarks are scientific excellence and technological innovation to advance the understanding of nature and to engage people of all ages in learning about birds and protecting the planet. Visit **www.birds.cornell.edu**.

Download the Free
Merlin Bird ID App!
From the Cornell Lab of Ornithology
Merlin.AllAboutBirds.org

The **Cornell** Lab of Ornithology
Merlin®

978-1-58355-936-9
ISBN
$7.95 U.S.
$9.95 CAN
50795
9 781583 559369

UPC
8 84682 00660 0
1 0 9 8 7 6 5 4 3 2 1
Made in the USA

3

The**Cornell**Lab of Ornithology

WATERFOWL ID SERIES

3 Sea Ducks & Others

BY KEVIN J. MCGOWAN

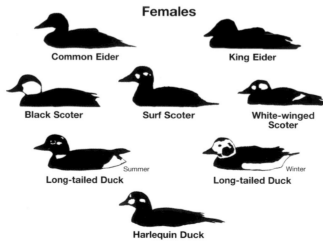

Download the free
Merlin® Bird ID app

Where's the WHITE?

T0124031

SEA DUCKS

These ducks are found almost exclusively on large bodies of water in winter, especially along the coasts. Most are heavy-bodied ducks that show bright patterns of black and white in the males. The females are dull and more camouflaged.

Silhouette ID – Where is the White?

Just spotting the white markings on a bird can lead to a quick and easy identification.

Males

Common Eider   **King Eider**

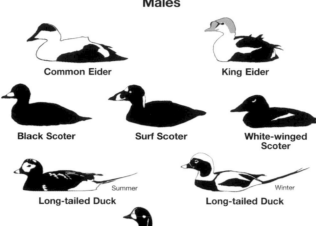

Black Scoter   **Surf Scoter**   **White-winged Scoter**

Summer   Winter
Long-tailed Duck   **Long-tailed Duck**

Harlequin Duck

Females

Common Eider   **King Eider**

Black Scoter   **Surf Scoter**   **White-winged Scoter**

Summer   Winter
Long-tailed Duck   **Long-tailed Duck**

Harlequin Duck

SEA DUCKS

Common Eider
Somateria mollissima
19.7-28 in. (50-71 cm)
Distinctive wedge-shaped head with long bill. Male bold black and white, with black cap, white back, and black undersides. Female brown with black barring; best distinguished by head shape.

King Eider
Somateria spectabilis
18.5-25.2 in. (47-64 cm)
Male with black body and white chest, with brightly colored head, including red-orange bill with large orange knob outlined with black. Female brown with black barring. Her bill is gray, and not enlarged.

Black Scoter
Melanitta americana
16.9-20.9 in. (43-53 cm)
Male all black with swollen yellow or orange knob at base of bill. Female blackish with large whitish face patch. No white in wings.

Surf Scoter
Melanitta perspicillata
18.9-23.6 in. (48-60 cm)
Male black with one or two white patches on head; bill colored white and orange. Female dark with whitish patches on face. No white in wings.

SEA DUCKS

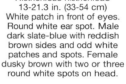
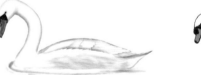

White-winged Scoter
Melanitta deglandi
18.9-22.8 in. (48-58 cm)
Black with white patch in wings. Male with comma-shaped white patch around eyes. Female with whitish patches on face. White in wings can be hidden while swimming.

Harlequin Duck
Histrionicus histrionicus
13-21.3 in. (33-54 cm)
White patch in front of eyes. Round white ear spot. Male dark slate-blue with reddish brown sides and odd white patches and spots. Female dusky brown with two or three round white spots on head.

Long-tailed Duck
Clangula hyemalis
15-22.8 in. (38-58 cm)
Mostly black-and-white plumage, varying throughout year; whitest in winter. Always with all-black wings, dark chest, and white belly. Male has long tail feathers and neat markings. Female has short tail and smudgy patterns.

Summer

Summer   Summer

Winter   Winter

SWANS

Swans are the largest of the waterfowl, with long necks, small heads, and large bodies. Two species are native to North America, and another is widely introduced. Sexes look alike.

Tundra Swan
Cygnus columbianus
47.2-57.9 in. (120-147 cm)
Large, all-white waterfowl. Long, straight neck. Black bill and face. Yellow spot in front of eye; may be large, small or absent. Black on face narrows in front of eyes.

Trumpeter Swan
Cygnus buccinator
54.3-62.2 in. (138-158 cm)
Large, all-white waterfowl. Long, straight neck. Black bill and face. Pink along edge of bill, but no yellow on face. Eye hidden in black on face.

Mute Swan
Cygnus olor
50-59.8 in. (127-152 cm)
Large, all-white waterfowl. Long, curved neck. Orange bill with black knob. Black face. Legs black. Neck held in curve. Introduced from Eurasia; may be found in parks.

GEESE

Geese are large waterfowl, usually larger than ducks. They travel in large flocks. Geese forage in shallow water by tipping-up, and on land they graze on grass or find food in agricultural fields. They don't show much color, but have distinctive patterns of dark and white. Sexes look alike.

Canada Goose
Branta canadensis
29.9-43.3 in. (76-110 cm)
Black head with white patch on chin and cheek; long, black neck, brown body. White rump and undertail. Black legs and tail.

Cackling Goose
Branta hutchinsii
21.7-29.5 in. (55-75 cm)
Small goose. Black head with white patch on chin and cheek, long, black neck, brown body. White rump and undertail. Black legs and tail. Bill small and triangular.

Brant
Branta bernicla
22-26 in. (56-66 cm)
Black head, neck and chest. White, partly broken collar on neck, white streaks on side, and white undertail. Restricted to coasts. Atlantic brant have light gray bellies, Pacific brant have black bellies.

GEESE

White Morph

Blue Morph

Snow Goose
Anser caerulescens
27.2-32.7 in. (69-83 cm)
Two color forms. White morph: white all over, except for black wingtips. Blue morph: white head and front of neck, body dark gray-brown, rump and part of wings pale gray; belly may be white or dark. Both with black patch on edges of pink bill.

Ross's Goose
Anser rossii
22.4-25.2 in. (57-64 cm)
Tiny white goose with black wingtips. Pink bill short and triangular. Greenish warty patch at base of bill. Lacks black along bill edge. A blue morph is very rare.

Greater White-fronted Goose
Anser albifrons
25.2-31.9 in. (64-81 cm)
Forehead and base of bill white. Body gray-brown, with white rump and undertail. Thin white stripe along side. Belly speckled with black. Legs orange. Bill pinkish to orange.

LOONS

Loons are large swimming birds with long bodies that slope to the rear and stout, straight, pointed bills. They sit low on the water and dive after fish, remaining under water for long periods of time. The sexes look alike, but loons change from brilliant breeding plumage with spots, lines, and patches of color to duller, more black-and-white winter plumage.

Winter

Summer

Common Loon
Gavia immer
26-35.8 in. (66-91 cm)
In summer, mostly black with white chest; thin black stripes on neck and sides, spots on back. In winter mostly white underside, darker back and neck. Indistinct white wedge on neck. White crescents around eye. Long, thick, pointed bill.

Summer

Winter

Red-throated Loon
Gavia stellata
20.9-27.2 in. (53-69 cm)
Dark gray with a red throat in summer. Paler gray back and white underside in winter; eye surrounded by white face. Thin, pointed bill tilted slightly upward.

Summer

Winter

Pacific Loon
Gavia pacifica
22.8-29.1 in. (58-74 cm)
In summer back black with white spots, gray crown and nape, dark throat with iridescent purple patch. In winter dark above and white below. Eye half in dark of face, half in white. Rounded head. May show dark chin strap.

Summer

Winter

GREBES

Grebes are small-to-medium-sized swimming birds with straight, pointed bills and compact bodies. They dive under water after food. The sexes look alike, but grebes change from colorful breeding plumage to duller, more black-and-white winter plumage.

Winter

Summer

Eared Grebe
Podiceps nigricollis
11.8-13.8 in. (30-35 cm)
Small head and short, thin, dark bill. In summer, black with golden ear tufts. In winter, black, white, and gray, with white ear patch. Front of neck grayish. Chin white. Rear end rather high off water.

Summer

Winter

Horned Grebe
Podiceps auritus
12.2-15 in. (31-38 cm)
Small head and short, thin bill. Reddish neck, black cheek, and yellow tuft behind eye in summer. Black and white in winter. Broad white cheek marks with dark cap. Whitish neck may be dusky at base. Rump near water.

Summer

Winter

Red-necked Grebe
Podiceps grisegena
16.9-22 in. (43-56 cm)
Bill large, straight, and sharp. In summer has rufous red neck, white cheeks, and black cap. In winter has gray neck and whitish cheeks.

Summer

Winter

GREBES

Winter

Summer

Pied-billed Grebe
Podilymbus podiceps
11.8-15 in. (30-38 cm)
Brown with tufted, whitish rump. Bill short and thick; pale, with black ring around it in summer, plain tan in winter. Throat black in summer, whitish in winter.

Summer

Winter

Least Grebe
Tachybaptus dominicus
8.7-10.6 in. (22-27 cm)
Range restricted to southern Texas and southward. Small, with sooty-colored head and body. Thin, dark bill. Yellow eyes. Tail end often high off water, showing fluffy white feathers. Throat white in winter, dark in summer.

Summer

Winter

Western Grebe
Aechmophorus occidentalis
21.7-29.5 in. (55-75 cm)
Long, thin neck. Long, thin bill. Black back and face. Eye in dark of face. White neck and underside, with black extending down shoulder and sides.

Clark's Grebe
Aechmophorus clarkii
21.7-29.5 in. (55-75 cm)
Long, thin neck. Long, thin bill. Black back and cap. Eye in white of face. White face neck, sides, and underside.

COOTS, CORMORANTS, ETC.

Other types of birds swim and might resemble ducks. Coots and gallinules are the swimming members of the otherwise wading rail family. Cormorants are larger birds that often are seen perched out of water, frequently spreading their wings to dry.

American Coot
Fulica americana
15.4-16.9 in. (39-43 cm)
Blackish all over, with darker head, and short, stout white bill. White patch beneath the short tail is used for display. Tail held close to water. Sexes look alike.

Common Gallinule
Gallinula galeata
12.6-13.8 in. (32-35 cm)
Bright red forehead shield and small chicken-like red bill with a yellow tip. Dark body with a thin, broken white line along side and some white under tail. Tail and wingtips usually held raised off the water. Sexes look alike.

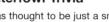

Double-crested Cormorant
Phalacrocorax auritus
27.6-35.4 in. (70-90 cm)
The most numerous and widespread North American cormorant. Occurs in large numbers inland as well as on the coast. Large and dark, with long body and long neck. Medium-sized bill is blunt or hooked at tip. Sexes look alike, but juveniles have pale upper breast and throat, and whitish to dusky chest.

Waterfowl Trivia

- The cackling goose was thought to be just a small form of the Canada goose, but research on their genetics showed they are quite different.
- Western and Clark's grebes were long thought to be the same species until careful observation showed that they did not breed with each other, even though they nested on the same ponds.
- Despite being fine swimmers, coots and grebes don't have webbed toes the way ducks do. Instead, they have several expanded lobes on each of their toes that act as paddles.
- The common gallinule is a fine swimmer, but it has long, thin toes without any kind of webbing.